IT'S ALL ABOUT TIME

Why the natural world is not the virtual

World that we have created

Published by Lulu.com 2012

Copyright 2012 by David Halsey

ISBN 978-1-105-57789-5

CONTENTS

PREFACE

Many of the philosophies that shape our societal characteristics can be traced back 2600 years to the Greeks. That was the time when individuals started questioning whether or not the "gods" were responsible for controlling nature's events. They had a "god" for every natural event that occurred. Many societies have risen and fallen since that time; however, the more each of those societal perceptions altered human thoughts about natural reality, the more the interpretation of societal behavior as a whole has remained the same or similar. From this knowledge, one could conclude that the behavioral attitudes of societies as a whole are "going forward to the past".

The culprit could be that the genetic capacity or make up of the preprogrammed Homo sapien brain which has not appeared to have evolved for millennia. Maybe Nature has concluded that there is no need for the brain to evolve beyond its original form since there will never be a time when Homo sapien will tax the basic survival and propagation instincts that control the human entity.

Since the beginning of the Homo sapien species, there has been an influx of knowledge about natural phenomena. Some individuals know more about science that defines Nature today than they knew yesterday but does society as a whole know more or care to know more about Nature today than yesterday?

So it can be written that in 2600 years our society has merely traded "Greek gods" for 21st century "causes" or "politically correct" attitudes.

My objective is to create a minute crack in the seam of understanding how Nature really works. Only time will tell! Thus *It's All About Time*!

David Halsey

It's All About Time

INTRODUCTION

"We not only believe what we see: to some extent we see what we believe."(Richard Gregory (1923-2010)
"Although we have no rational grounds for believing in an objective reality, we also have no choice but to act as if it's true." David Hume (1711-1776)

"...there is no measurable curvature in the Universe today at all." Sean Carroll

This book is based upon the axiom that Nature is what it is, not otherwise. This is the paradigm that I build my concept of reality upon. Nature seems to take the simplistic side of any argument, thus my objective is to look for the simplicity of reality as defined by Nature, not to describe the virtual world that society has created. My goal is to imagine a natural world that can be described within the realm of acceptable scientific knowledge and interpretation of observable "ground-truth" data.

Time is presented as the major component that has shaped the Universe and all its elements into an illusion that we call reality. Theoretical physics and cosmology are the fabric of this book, but chemistry and biology are threaded

1

throughout. Explored chronologically are the implications that time is the glue that binds the microcosm to the macrocosm-from infinity to infinity. This includes how time distinguishes matter from an atom to a galaxy; from a Black Hole (BH) to an atom's nucleus; from a definition of "Dark Matter" to the existence of a collection of atoms defined as "living" entities. Also included is a discussion of how the entropy of the Universe, a closed system, affects our reality. Evidence is beginning to emerge that implies that the laws of nature, that we have devised, are changing with time. This adds to the notion that the more we know/learn about Nature the more we know that we don't know!

We begin with an introduction describing many truth-seeking questions that are as old as men who were curious about how Nature works. Is the past real? Is the future real? What is reality? How does space-time define Nature vs reality? What is time? Why time? How did we get here in the first place?

Regardless of one's thoughts, beliefs or theories as to how life got to earth and evolved into all of Earth's critters, including us, the fact remains, that it did. Thus we know that the Universe is *"life friendly"* because we exist! Some entities ask the question: "Was the Universe designed for life?" My response is no because it's vice versa.

Also included is a discussion of the conflicts between the "virtual worlds" that we have created versus the actual world that Nature has provided. For example, all the measurements that we use to keep our bearings are made up. The constellations exist only in our imagination. We modeled the inside of an atom after a miniature solar system yet we suspect that an atom's insides are much different.

It's All About Time

The *Periodic Table* is a tabular display of the 118 known chemical elements organized by selected properties of what we believe about their atomic structures. Thus elements in the *Table* are organized based upon how they perform not how they were formed. Presented is a theory showing how time inside the nucleus distinguishes how each atom appears to be similar but not the same. This difference is in the state of matter, energy vs mass. All atoms appear to contain the same amount of matter as the hydrogen atom; it is in a different ratio of energy vs mass that identifies each element.

Time's relationship with gravity is discussed as to how time is the single ingredient that transforms the four fundamental natural forces into one unified force, gravity. Included is how time shapes the space continuum to prevent everything from happening at once.

Chapter 4 shows how galaxies are a clock in motion. From their Black Hole nucleus to the outer most star time increases. Included is the mechanism that allows galaxy matter to rotate in a synchronized fashion.

The Universe's natural form must be a logarithmic spiral that allowed for early inflation and growth, yet present an appearance of being flat. Explained is how time increased from near zero at the Big Bang to present day rate, logarithmically, which causes an error in the calculation of the Universe's age.

The speed of light is time's yard stick, therefore varies in every time zone. This phenomenon is discussed at length.

Everything in the Universe seeks a lower energy state. This process is called entropy. Since there appears only to be non-reversible processes in the Universe, time has only one direction.

Beginning with Chapter 8, we discuss how time has allowed for the creation of the elements that produced life. The question is the Universe designed to perpetuate life based upon the design criteria recorded in the DNA helix? The fact that microbes can survive in 'space' without the need to replicate and process nutrients for energy for millions of years is discussed and proven, in relation to how the flow of time affected this process is discussed. The proof that comets ('snowballs" from the Jupiter region) which carry the water molecule that is identical to the one found on Earth plus bring carbon dioxide (Earth's source of carbon and breathable oxygen) as well as microbes to Earth is discussed. The fact that evolution on Earth produced Homo sapiens who in time created this virtual world is discussed from the Greeks (5th century B.C.) to the 21st century, e.g. - Science vs philosophy.

In time humans have devised societal rules and edicts that have nothing to do with natural survival and perpetuating the species. Today some believe that science has made philosophy less important, or even dead, simply because philosophy has not kept up with modern developments in science, particularly physics.[1] It is a fact that some spiritual "believers" have not left the 6th century; other cult groups still reflect 17th century philosophy; and still others that maybe have journeyed into the 19th century but it appears that no philosophy has made it to the 21st century.

Protagoras (ca 490 B.C) was a pre-Socratic philosopher. He expressed his views on religion which seem to be still the most rational quote concerning the subject: "Concerning the gods, I have no means of knowing whether they exist or not, nor of what form they are; for there are many obstacles to such knowledge, including the obscurity of the subject and the shortness of human life."

---

[1] Hawkins, Stephen, & Mlodinow, Leonard, "The Grand Design", Bantam, 2010

It's All About Time

Arthur Koestler in his book, "The Sleepwalkers"[2], stated that during the Middle Ages (a historical period following the Iron Age fully underway by the 5th century and lasting to the 15th century) Christianity saved Europe from lapsing into barbarism. He also said that the catastrophic conditions of the age and the climate of despair prevented Europe from evolving a balanced, integrated, evolutionary view of the Universe and of man's role in it.

Our psyche is trapped inside a mental prison and our only communication with the outside world is through our senses. How we receive impressions of the world external to our bodies depends upon the quantity and quality of our senses, i.e.-sight, hearing, taste, touch and smell. This is why 10 people witnessing the same event will describe the event 10 different ways. We are ALL "wired" differently. If you don't comprehend this concept then I have just two words to summarize: *Helen Keller.*

Finally Earth's climate(s) are discussed in detail about "how" the virtual world describes what is happening rather than "why" the climate is what it is in the natural world in which it exists. From dust to ice; from the ocean thermal currents to why carbon dioxide WILL NOT rise in the air column; to how carbon and oxygen has sustained life; to the effect of comet bombardment for millions of years are discussed from the microclimate to the regional climates that cover the globe. Why Antarctica is Earth's thermostat and why Earth's climate default position has always been "ice" is presented.

---

[2] Penquin, 1959

It's All About Time

The Epilogue summarizes the contents of the book as to the "WHY" question, which are the answers that most scientists believe will take science to its knowledgeable limits.

The study of science simply means the collection of information. As we discover Nature's way of doing business we add to the whole depository of mankind's knowledge. There is one caveat: The "truth" means different things to different people. The validity of that truth is a function of the quality and quantity of each individual's senses.

The Greek philosopher Thales (624 BC – c. 546 BC), who some believe to have begun the Greek philosophy movement, challenged the Greek Gods about who controlled the "natural" forces of the known world. Western civilization is built upon the knowledge and traditions of the ancient Greeks. They were the first to observe the world with questions about where did everything come from; where does everything go; why does everything work? Thales began the search for the laws of nature by separating the quest from the gods of mythology. These thoughts were the birth of scientific study about the world.

Pythagoreanism[3] followers believed that the Sun was the center of the Universe, but for more than 1700 years Aristotle's[4] views on the physical sciences

---

[3] Pythagoras of Samos (c. 570–c. 495 BC) was an Ionian Greek philosopher, mathematician, and founder of the religious movement called Pythagoreanism.
[4] Aristotle (384 BC – 322 BC) [1] was a Greek philosopher, a student of Plato and teacher of Alexander the Great.

profoundly shaped medieval scholarship. His influence extended well into the Renaissance, although it was ultimately replaced by Newtonian[5] physics.

My research and conclusions are a combination of my imagination and the knowledge that I have gleaned from a lifetime studying engineering, natural science and physics. They are written in terms that I care to leave for future readers about my past as I continue travelling through time toward that "real" event in the future, my death moment.

---

[5] Sir Isaac Newton PRS (4 January 1643 – 31 March 1727 [OS: 25 December 1642 – 20 March 1727])[1] was an English physicist, mathematician, astronomer, natural philosopher, alchemist, and theologian.

It's All About Time

CHAPTER 1
TIME-THE CATALYST

"Time is Nature's way of keeping everything from happening at once." John Archbold Wheeler

"The reason that time has a direction is because the Universe is full of irreversible processes." Sean Carroll[6]

This fleeting instant, we term the present is the boundary that separates the future from the past. Time is not continuous but flows in increments that are so minute one cannot readily imagine. The dimension of this lull we call the present is the only moment when time stands still. Thus the "present" is of little significant to reality, it's simply a boundary between the past and future.

The two questions that have bugged scientists since the "dawn of science" are: (1) How did life start and (2) What is consciousness? Each leads to: What are we? How did we get to this place? Are we alone? Etc., etc., etc...

I have analyzed and researched all the data that I could mentally ingest in the last 60 years and I have arrived at a conclusion that many before me have made. "We don't even know what is possible to know!" The passage of time appears to be continuous in the world that we live in. Its measurement always depends upon one physical system relative to another. For instance, how many times does a clock hand rotate relative to the rotation of the Earth? These events are not related nor does one cause the other. Many time measuring devices do not reflect the natural world. Since the bits of time are so small that

---

[6] "From Eternity to Here", Dutton, 2010

we cannot measure them with today's technology, we have attached an arbitrary rate of one second per second to the flow of time. We record the rotation of the Earth as a day, divided into 24 hours; the time it takes the Earth to rotate around the Sun as a year yet neither measurement is precise thus the need for a leap year every four years to reset and make the system work. Furthermore day and year measurements are not relative nor do they depend upon one another in any way. Thus time measurements in this world that we have created are both relative and each arbitrary.

If only we could anticipate the future. At family reunions I spend time catching up on the lives of kin folk that I have not seen or heard from in a year, and sometimes longer. Many times as we get older I am saying goodbye to kin without knowing it. In every case there are many family related thoughts that go unsaid. It is a reminder that for the living time is the cruelest dimension of all yet without time there would be no past or future. Thus without time there would be no memories, no good old days to reminiscent about, no myths to pass on from generation-to-generation.

Time is the integral part that defines our daily lives. We describe our activities around time. For instance what did we do last week or month or what are we planning for tomorrow all involve the past or future time. We identify and celebrate special events by "moments" in time but we despise its passing. Our hopes and fears are labeled in terms of time. "Do you remember the time that so and so did...?"Time is our fourth dimension. Length, width and height make an object appear solid but time makes it real in appearance. For instance you are sitting in a specific desk in a particular room of a school building, which is described by an address and room number at a certain time of day, say two o'clock in the afternoon. At three o'clock you are still in the same seat but the past has eaten up another hour of the future. At some later[7] time you have moved on therefore you are no longer in the chair, so time defines your location, past, present or future.

We know that the arrow of time always moves forward. That is to say that the past always eats into the future at a rate of one second per second. What you see when you are talking to a person is that person in the past. The time it takes for the light image and the sound to travel from that person to you puts them in your past. Thus nothing you are aware of is in your present time. By definition the "present" is impossible. The adage that we are always living in the past is absolutely true; otherwise it would be impossible to communicate.

You are driving down a country road. When you look through the windshield you are looking at where you will be in a minute or so. Thus you are looking into the future. On the other hand if you look through your rearview mirror you see what you just passed a few moments before. This is your immediate past. In this scenario you can see the future and the past so they both must be real.

All three time dimensions, past, present and future have always existed. The definition of time demands it. For example your birth date is in the past and since you exist the date must be real. Since you will die in the future your death

---

[7] I never understood why the term "later" should define a future event.

date does exist; so do your future birthdays, anniversaries, etc. You are moving toward these dates at a rate defined by time. So your future does exist. These points in time have always existed. On the contrary there is a quote from Einstein: "The past, present and future are illusions". The definition of "illusion" could be "a figment of our imagination".

Every entity in space–time has a "base" from which time can only move forward to the future. This eliminates the possibility for any paradoxes. Some physicists believe that the past and future exist in a space-time continuum. Simply stated this means that the four dimensions, length, width, height and time define an event such as your birth and its position in a space-time continuum. This is to say that your birth's past still exists somewhere in the space-time continuum. If so, it would be located at a distance from you equivalent to your age times the speed of light.

Since we don't even know what is possible to know there is evidence that's beginning to emerge which implies the laws of nature, which we have discovered and defined by our technological theories, are changing with time. Since the concept of time is what distances eternity from the beginning these theories will no doubt change as we gain a better understanding of the natural world. I believe the flow of time is the catalyst that takes reality from the present to the future and provides the only avenue in which the past can be quantified. Since time is merely the rate of change from one spatial configuration to another, it is the fourth dimension.

CHAPTER 2
ATOMS ARE IDENTIFIED BY AN INTERNAL CLOCK

"By convention there is color, by convention sweetness, by convention bitterness, but in reality there are atoms and the void,' announced Democritus. The universe consists only of atoms and the void; all else is opinion and illusion. If the soul exists, it also consists of atoms."

The secret to the definition of time could be revealed if we could discover how the atom was originally formed. The collecting, combining and "jump" starting particles to function as an atom is the question that needs defined. All atoms, born now in an expanding Universe, and in the beginning, are hydrogen atoms which appear to be the basis for ALL atoms.

Atoms are nature's storehouse for energy. They are the perfect battery. It is believed that 99.99% of an atom's mass is stored in its nucleus. Because of this stored energy the time zone within the nucleus is slowed to near zero because of the strong force. This puts the atom in a time warp with its outside environment. This could be one reason that a hydrogen atom, formed just after the Big Bang (BB), continues to exist in its original state after billion of years. Within this atom's nucleus a second is an eternity!

All atoms after hydrogen are simply modification of the hydrogen atom. Within the hydrogen atom the nucleus' matter is mostly in the form of energy because of a near zero time rate that exist in the nucleus as a result of being formed at the time of the BB. As the hydrogen atoms internal clock is altered by a cataclysmic event, i.e. a supernova or simply a star burning through its life cycle, more energy from its nucleus is converted into mass (particles). The "changed" atom takes on a different set of characteristics. Each time the new atom is subjected to this violence its mass grows and energy decreases as it becomes still a different atom. Supernovae are important in the nucleosynthesis of heavy elements. The explosions take a few tens of milliseconds creating selenium, copper, zinc and all the heavy elements that are found on Earth, and in our bodies. Synthetically, we have created elements that cannot survive in our time zone. Recently (2009) an element number 112 was created which is 277 times heavier than hydrogen. No. 112 resides on top of the Periodic Chart. This element was formed when zinc and lead were fused. This so called super heavy element existed for only a tiny fraction of a second before decaying, radioactively, into the elements that can exist within our time rate.

Hydrogen appears to be the "lightest" of all atoms because the matter of its nucleus exists almost entirely in the form of energy as oppose to mass. Thus the hydrogen atom has the greatest potential to be the most powerful of all atoms. Why gravity has the appearance of having a lesser effect upon energy compared to mass, within the atom's nucleus, is probably because of the strong force, a gravity component that bonds the nucleus.

Fusion[8] allows the harvesting of energy from an atom. This appears to occur when we manipulate the internal clock of certain atoms. An example is the

---

[8] Nuclear fusion is the process by which two or more atomic nuclei join together, or "fuse", to form a single heavier nucleus. This is usually accompanied by the release or absorption of large quantities of energy. Fusion is the process that powers active stars, the hydrogen bomb and experimental devices examining fusion power for electrical generation.

energy released by causing a nuclear explosion by abruptly increasing the atoms internal clock, for an instant, to one second per second, the present time rate.

According to Fuller chemistry[9] is seen as "applied physics", whereas chemists claim their science is of things perceived by the human senses. From a physicists point of view it's conceivable that all atoms are identical in total "matter" but the internal structures of their nuclei are altered by virtue of the distribution of energy vs. mass. The difference in their respective uniqueness is the internal clocks-that-tick within the nucleus?

From a chemists point of view the *Periodic Table* is an orderly arrangement of known elements. It doesn't explain how the elements were created or what makes them similar but not the same. As we move up the *Periodic Table*[10], from hydrogen, the infinitesimal increases in the flow of time, within each atom's nucleus, cause the release of energy that allows certain particles, equivalent to that specific time, to be manifested, thus modifying the internal structure of the original hydrogen atom, creating a "new" atom. The new atom's atomic weight would be heavier because of the number of particles appearing as mass within the atom. Energy appears to not be affected by gravity thus it presents no

---

[9] Fuller, Steve; Kuhn, Thomas, "A Philosophical History for Our Times", University of Chicago Press, 2000

[10] In 1869, a Russian chemist Dmitri Mendeleev published the "Periodic Table". All known elements were arranged in order of their atomic weights and similar properties. The modern Table contains the atomic number (number of protons in the nucleus), element symbol, element name and atomic weight (approximate number of protons and neutrons in the nucleus).

weight. We must conclude that elements that are created by synthetic methods that exist only for an instant, have a nucleus time rate differential that cannot survive in our time of one second per second. As universal time increases in the future, which it has done since the BB and must increase as the Universe expands; these elements will be stable in some future time zone. Nature has accomplished this same task of releasing energy in a seemingly less violent fashion by what we call the "weak force". Particles we term as radiation appear to be in sync with one second per second time passage and are released at certain half-life rates. These particles were trapped in the atoms nucleus eons ago, finally traversing through the atom's time zone into present day. An analogy would be a light photon taking millions of years to travel from the Sun's center to its surface before making the eight (8) minute trip to Earth. The length of time is different because the time zone at the Sun's center approaches zero as a limit and on Earth its one second per second.

As a black hole (BH) ingests an atom, the nucleus (held intact by the strong force) is squeezed into the singularity while the electron cloud (held together by a weaker force, the electromagnetic force), containing less than 0.001% of the atom's mass, is expelled in a tightly focused beam or jet along the axis of rotation of the BH's singularity. The size and velocity of these jets indicate the quantity of mass being ingested and the residual energy of the BH.

It is generally accepted that the gravity inside a BH is so strong that light photons, which have mass, cannot escape. There is at least one other logical reason that prevents light from escaping a BH: There are no particles intact that can produce light. Time flow near the BH's singularity approaches zero as a limit therefore one second is an eternity.

BHs appear to be nature's local recycling center as mass is converted back into energy.

It's All About Time

CHAPTER 3
GRAVITY AND TIME

Gravity is only the bark of wisdom's tree but it preserves it" Confucius (551-479 BC)

The magnitude of gravity is dependent upon the location and density of the mass producing the gravity. Also, the gravitational field itself has mass and that mass, contributes to the total force exerted by the gravity field.   Gravity has an inverse effect on the flow of time. As gravity increases time slows down. As previously stated within a BH time approaches zero as a limit. There appears to be one exception to this rule. Gravity does not react with matter in the energy state, but only in the mass state. At least this appears to be true for the Strong Force.

Time is the single ingredient that transforms the four known fundamental natural forces into one unified force, gravity[11]. Here's how: within an atom the 'strong force' is present because the local time zone in the nucleus is near zero. If you apply Newton's gravitation equation[12] to the nucleus of an atom, replacing the 'gravitational constant' with a time of zero +, then gravity and the strong force appear to be the same. If the rate of time is increased then the weak and

---

[11] It's conceivable that we have been seduced by gravity and ultimately the strong force is the "unified force." It appears that the nuclei of atoms, BHs and the Singularity are intact because of the strong force; therefore gravity could be an aberration?

[12] $F=Gm_1m_2/d_2$; G=rate of time flow instead of the gravitational constant; m_1=nucleus (mass); m_2=to electron (mass) and d_2= distance between the nucleus and electron mass. Therefore the force of attraction (F) = the strong force which = gravity. $F= (0+) m_1m_2/ (0+) 2$.

electromagnetic forces are in fact just simply different forms of gravity within different time zones. If time flow is increased to the present then gravity is as we experience it.

The energy stored in atoms, neutron stars, BHs and the Singularity, appears to decrease time within their respective time zones in proportional to their respective matter. Within the latter three, gravity appears to be the only natural force present. Take the BH for instance; the weak, strong and electromagnetic forces only exist, in time, up to the BH's event horizon[13]. At this point the decrease in time appears to convert all the natural forces into one force, gravity.

Somehow gravity and time are intertwined to the point where one is the manifestation of the other. Gravity and all its sub-forces binds the Universe whereas time is the medium in which energy can convert into mass.

The "current grand unification theories" (GUT) of forces include the two nuclear forces, the strong and weak, and the electromagnetic force but not gravity.   The strong force is the force that binds together the particles in the atom's nuclei.   The weak nuclear force is responsible for radiation. The electromagnetic force only acts between charged particles and is the transmitter of information between molecules/cells, etc.   Electricity is a product of the electromagnetic force. While the gravitational force is the weakest of the four forces by many orders of magnitude, it interacts over huge distances between all particles.

---

[13] The event horizon is the threshold where nothing can escape.

It's All About Time

With an increase in temperature and pressure the two nuclear forces weaken, with the weak force petering out first. With these conditions the electromagnetic force grows stronger.

Gravity has an inverse effect on the flow of time. If you use a BH as a model the two nuclear forces dissipate before the BH's horizon is reached, whereas the electromagnetic and gravity increase in magnitude. It's conceivable (see footnote 14) yet not proven that we have been seduced by gravity and ultimately the electromagnetic force (the force that transmits information) is the "UFT." Just maybe all information is not lost in a BH but recorded by the electromagnetic force?

Gravity and time appear to be more intertwined than time and the electromagnetic force. Therefore gravity could be the manifestation of time.

It's All About Time

CHAPTER 4
GALAXIES-A CLOCK IN MOTION

"All galaxies are moving away from all the other galaxies, and each of them sees the same kind of behavior. It's almost as if galaxies aren't moving at all, but rather that the galaxies are staying put and space itself is expanding in between them." Sean Carroll

Black Holes are the nuclei upon which galaxies are formed. It's feasible to imagine that BHs were originally clumps of matter from the Singularity[14] that survived the inflation phase[15] of expansion and remained as energy islands that were so strong they were not converted into mass. These BHs 'seeds' became the nuclei that shepherded the mass into gas islands called nebulae, later galaxies, making star formation possible.

The ingestion of matter by the BH nucleus drives the rate and direction of rotation of a galaxy. The angular velocity curve for mass rotating around the nucleus, as you move away from its center, increases initially then becomes flat[16], creating a logarithmic fashioned curve. The angular velocity of the mass in the outer three-fourths+/- of a spiral galaxy is nearly constant. In a galaxy, as

---

[14] An infinitely dense point located at the center of a "black hole", where the current laws of physics no longer apply. A mathematical point at which space and time are infinitely distorted.
[15] The initial period of the beginning of the Universe expansion. Here is where time increased the fasted from near zero to what it is today, one second per second. The expansion followed a logarithmic scale.
[16] Rubin, V., Thonnard, N., Ford, Jr., W. K., "Rotational Properties of 21 Sc. Galaxies with a Large Range of Luminosites and Radii from NGC 4605 (R=4kpc) to UGC 2885 (R=122kpc)", Astrophysical Journal 238:471, 1980

opposed to the Solar System, the distribution of mass is such that the resultant forces, whether from gravity, space-time and/or a combination of both, are balanced, centrifugally, so that rotation scribes a near circle.   Stars/Solar Systems are like planetary gears rotating inside the larger (galaxy) gear. The gravitational 'family' effect of the BH nucleus is manifested through these 'planetary gears'. This allows the mass concentration to cling to its neighbor via a "gravity chain"; similar to a group of paper clips clinging to one another as they hang from a single pole magnet.

The differential of time from the galaxy's nucleus, at near zero, to the outer edges also contribute to the near synchronous rotation of all galactic mass. The stars and their applicable solar systems in each galaxy has a rotation that is vectored slightly inward as mass is being fed into their respective BH nucleuses.

The dwarf galaxy creation goes back to the early days of the Universe because of their massiveness. (I.e. dark matter) The dwarfs have lost most of their mass to their respective nucleus (BH), thus they have very little mass left with which to create stars.

In the 1980s the" Parkes All Sky Survey" (Parkes Radio Telescope) called HIPASS, scanned the southern sky and found hundreds of galaxies that were previously hidden behind the Milky Way, thus invisible to optical detection. Analysis of 2% of the data detected a large number of dwarf galaxies. Because so many exist the preliminary conclusion is that dwarf galaxies may contain the "missing mass", including the "dark mass" that some believe is needed to make the Universe's dynamics work. My belief is this mass is found in the respective galaxy's nuclei, a BH, of each galaxy in the form of energy.

Globular clusters do not have a preponderance of "dark matter" because most of their mass is visible. The nucleus appears to contain a relatively small mass that has not had sufficient time to ingest or organize the galaxy's mass to rotate. These are evidential signs that identify a young galaxy that contains a mass similar to a large galaxy but in a different state, e. g. - Visible vs. BH or mass vs. energy.

The fact that smaller galaxies are more numerous than large galaxies is an indication that the Universe is middle age. My calculations suggest something between 21.5 to 42.5 billion years (present time rate) old. At this age there is time for some galaxy's nucleus to ingest enough mass to grow into a substantial BH.

It's All About Time

CHAPTER 5
UNIVERSE-A LOGARITHMIC PROGRESSION

"If the whole universe has no meaning, we should never have found out that it has no meaning: just as, if there were no light in the universe and therefore no creatures with eyes, we should never know it was dark. Dark would be without meaning." C.S. Lewis (1869-1963)

Past   Present day

My Universe

Each compartment represents the flow of time which causes the Universe to (face of the shell at any past time) expand to present day size. The "face" is always flat! The conceptual diagram records the past history of the Universe as time follows *a logarithmic pattern that creates the future. The dimensions at the face are infinity in every direction.*

From the sensing capability of Homo sapiens, developed within the same environment that is being sensed, the most logical model for the Universe appears to be the **"Spiral of Theodore of Cyrene".** This natural form is called the Nautilus shell, which is a logarithmic spiral. If so, the spiral horn begins at the Singularity, the microcosm, which retains 99.99% of all energy, and spirals, infinitely outward, to create the macrocosm, our finite world. This design allow for a conservation of energy as the Universe expands. The face or cross section of the Nautilus is the visible Universe of today, thus the flat illusion. Scientists have looked back into the spiral for 13+ billion years, our clock but not the clock of the mass being observed. The farther back in time we observe the less mass we see, thus there is more invisible energy.

What we can agree on is the portion of the Universe that we can observe is less than a minute of the whole. There are hints that the observable Universe, the portion at the mouth (face) of the 'horn', is mass that has formed into spiral arms. We have observed what we call the "Great Wall" and other concentrations of mass to substantiate the spiral effect. In reality the Universe is so immense that we probably will never comprehend its form.

If time began at a near zero rate at the birth of the Universe and increased as the Universe expanded, its volume increase is 16 times faster than time measured (volume vs area). Yes the Universe is expanding but not at the rate that is observed but 16 times slower. The expansion rate appears to follow a logarithmic pattern, with infinite cycles, which appears to be a logical natural process.   The logarithmic concept would account for the inflation period of growth.

Is the energy throughout the Universe an information source?   Is the information about the Universe contained in matter that is in an energy state, which manifests into mass, with time, before entropy ultimately dominates as the current Universal cycle ends and another begins? The energy causes the Universe to expand as matter is created with the flow of time. Order/information is that the expansion/whole Universe looks the same from any direction before the future end game called entropy.

The Universal system, like everything else within its domain, is subject to the laws of thermodynamics.   Once the whole energy system runs low on information, it cannot balance its account with matter, thus the Universe will fall apart (entropy) and only the renewal cycle driven by the Singularity can put it all together again.

Every corner of the Universe, from the microcosm to the macrocosm, has its own local time. From the flow of this respective time comes the mass state of matter that we can use to define the motion of any entity in relation to its neighborhood. What Einstein called the inertial frame in his special relativity theory (SR) could be the smallest locale with respective time zones that we can visualize until we have gained enough knowledge to delve deeper into the quantum world[17]. For now these frames exist only as a way to identify the passage of time.

---

[17] It appears that all particles, regardless of size, are made of smaller particles. This is true to

It's All About Time

The first known thought of a cyclic Universe can be traced to the Greek philosopher Plotinus, born in Egypt circa 205 CE. He suggested that an "eternal cosmic cycle" would produce infinitude of identical universes. Since Nature has never been the same in any scenario, it's conceivable that the 'eternal return' could produce an 'eternal cosmic cycle'.

Within nature everything begins anew, progresses through a cycle of change and eventually the whole ceases to exist, terminating into the original incremental building blocks/particles. Our Universe appears to be no different as, with the flow of time, it appears to be progressing from one methodical state to another as it ages, in an inescapable sort of way.

---

the infinitum.

It's All About Time

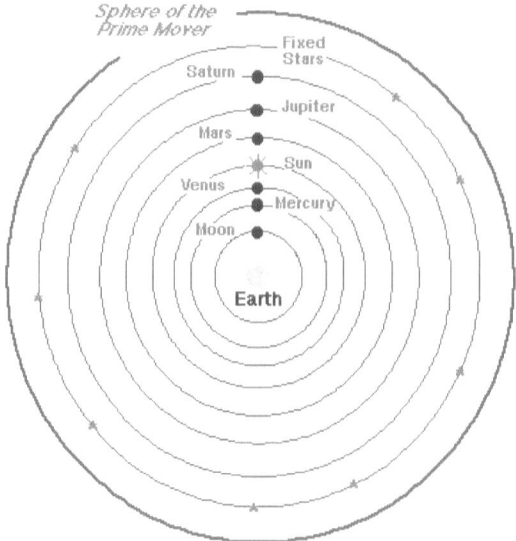

Aristotle's Universe

As the Universe ages, the rate of time increases as it's center of mass moves further away from the influence of the Singularity. Since a very small amount of mass can store an incredible magnitude of energy, (remember Einstein's formula $E=MC^2$ where E- energy is equal to M=mass times the speed of light squared) within any given time zone, this has driven the growth/expansion of the Universe. Since energy is invisible to our sensing devices it could conceivably be mistaken for the elusive 'dark matter'. There appears to be residual energies throughout the Universe that will, in time, convert to mass as the Universe expands.

In the beginning, less than one percent of the energy contained in the Singularity was dispersed at near infinite velocity and near an infinite energy state to expand into the Universe. From Newton's point of view this event was caused by an irresistible force meeting an immovable object creating what is described as the Big Bang (BB). The collapsing of the last Universe was the irresistible force and the immovable object was the Singularity.

When all the energy ejected from the Singularity has converted into mass via time it can be said the Universe has reached equilibrium. However, at this point in the age of our Universe there is a continuing conversion of mass back to its original energy state within the nuclei (BH) of each galaxy. This process was started the moment the first galaxy formed. By the end of the Universe's life some BH will have continued to grow by ingesting mass and merging with other BHs.

The moment when the original energy, which was expelled from the Singularity, is exhausted, time will have no other function and the 'Big Crunch" will commence. The moment time approaches zero as a limit all mass will become energy again, and visual reality will no longer exist. There will be only one force, gravity.

CHAPTER 6
SPEED OF LIGHT (SOL) TIME'S YARDSTICK

CHAPTER 6
SPEED OF LIGHT (SOL) TIME'S YARDSTICK

"Each ray of light moves in the coordinate system 'at rest' with the definite, constant velocity V independent of whether this ray of light is emitted by a body at rest or a body in motion." — Albert Einstein

Light is the medium that bring images to our sensor, the eye, which our brain interprets as reality. Even though the SOL is constant, travelling at a speed of 299,792,458 meters per second within our time zone, the distance it travels in a given time is dependent upon the time zone that the photons pass through.

When looking back into time we are looking at a slower time flow than present day value of one second per second. The light that we observed has travelled for nearly 15 billion years, our time standard, to reach our sensing device. During its journey the photons negotiated many time zones created by the space-time potholes and gravity traps along the way. Thus it's not too farfetched to say that instead of travelling for 15 billion years by our yardstick, in fact we are not sure when the light left its source! There must be a logarithmic effect in the progression of what we call the Universal clock.

Another observation that could be interpreted to support this concept is our observation of the stars orbiting the Milky Way's nucleus at a velocity of over one million miles per hour, relative to Earth's time. The stars lie very close to the BH, where time approaches zero, that drives the galaxy, therefore time has slowed and the effect on light photons leaving that region must be substantial. What would the effect be if this light was delayed for a few thousand million years Earth time? One other complication; what if these stars are found to be second or third generation stars? Besides being evidence that the nucleus is growing by ingesting mass, this would further complicate the age question.

CHAPTER 7
ENTROPY

"It was not easy for a person brought up in the ways of classical thermodynamics to come around to the idea that gain of entropy eventually is nothing more or less than loss of information". — Gilbert Newton Lewis

One reason that time has a direction is because there is a non-reversible process in the Universe called entropy. As the Universe ages entropy increases. When the energy (matter) from the BB reaches equilibrium entropy and time will decrease instantly causing the "big crunch" that will produce a new Universe.

It is a measure of disorder and is referred to as an "emergent property[18]" because it only emerges when a system becomes sufficiently complex. Since the Universe is a complex closed system, all natural processes are irreversible changes that can only continue because of time. In a closed system energy decreases as entropy increases with time. The natural processes that define matter appear to always conform to the laws of thermodynamics. These processes are irreversible, as far as we can surmise based on current knowledge.

BHs are the penultimate form of entropy compared to the ultimate form, the Singularity. Once any object is in the zone of influence of a BH, entropy increases until the object reaches the horizon or point of no return. This zone represents the maximum entropy. After the horizon point entropy begins to decrease rapidly as the object is reduced to its minimum particles that join the

---

[18] Primack, J. R., Abrams, N. E., "The View from the Center of the Universe", Riverhead Books, 2006

Singularity. Once inside the BH entropy is at its lowest. If the BH extracts radiation then entropy does not decrease to zero. The demise of the BH creates entropy which may rise as the BH evaporates.

Once the increase in time has converted all energy into mass, which has caused the Universe to "grow" to its ultimate size in order to contain the mass, entropy will have almost maximized. At this point in time energy in this closed system will be spread so thin, since most of it has been converted into mass, that the Universe should cease expansion. My sense is that a state of equilibrium will exist for a few billion (current time base) years followed by a contraction phase which will occur at the speed of light relative to the existing time. Remembering that nearly all the maximum energy imaginable still resides in the Singularity, maximum entropy will occur when all energy is returned to same.

Another feasible scenario relates to the flow of time. Since time appears to be the catalyst, or by-product, of a conversion of energy into mass, what would happen if time simply stopped, became zero? Would all mass suddenly revert to energy? If so, the visible Universe would be gone in an instant!

Throughout each Universe the final repository of information is entropy. If there are cyclic Universes then entropy is also cyclic. Thus, the second law of thermodynamics only applies to each specific Universe from birth to death.

Back on Earth everything that was "alive" dies and, with time, breaks down to dust, or to the elements called atoms. Here entropy is at its maximum and, in most instances, begins anew as the elements are ingested into another living entity. This cycle has continued on Earth since life first appeared and continues to the present.

Everything in Nature seeks a state of lowest energy. That is why water flows downhill, we turn to dust after we die, and the weight of snow pack causes avalanches. This hypothesis is true for all matter throughout the Universe. Even the gravity force that appears to make this all happen will in time become zero as its creator, mass, disperses because of entropy.

It's All About Time

CHAPTER 8
HOW LIFE GOT TO EARTH

"In three words I can sum up everything I've learned about life: it goes on." **Robert Frost**

The elements[19] to sustain life are continuing to be created in the evolution of the stars. It all begins with the simplest and oldest element, hydrogen, which is the fuel of stars. From this basic element, every other element that forms our body is manufactured as "star stuff". So the question begs: "Was the Universe designed to perpetuate life based upon the design criteria recorded in the DNA helix?"

One could begin to wonder if the complex DNA helix was in fact "made" within the lifetime of the Universe. We'll never know. However, what appears real, besides the complex magnitude of life on this oasis called Earth, are the "conveyors of life", comets, which exist throughout the realm of the visible Universe. Comets are made of water ice and most agree that a search for life is a search for water. As a matter of fact, as you will learn from future chapters that comets provided the ingredients to support the environmental conditions that made life on earth possible!

Will we eventually accept the fact that we are not alone?   Earth's population of life may be the only atypical copies in the Universe but there are likely forms of life in other worlds that have evolved by adapting to the environments of their "planet".

---

[19] Every element except hydrogen listed in the Periodic Table, up to Iron, was created in the life cycle of a star including the nova (star explosion).

It's All About Time

The microbes that NASA found in California's Mono Lake (2010) appear to have evolved by substituting arsenic for phosphorous as a nutrient to survive. This is one more clue that life is very resilient. The more knowledge we gain about the "particles of life" the more we are led to conclude that the Universe was designed to accommodate and sustain life. This chapter presents some of these arguments.

We have said that the laws of physics are obviously friendly to life because we exist! From the evolution of stars come the chemical elements that make up the atoms which have come alive. Every element, besides hydrogen, that forms our bodies was created within the life cycle of a star. This part we know. The element hydrogen, with a trace of helium has existed since the beginning of time. Currently hydrogen and traces of helium make up 99.9% of all visible matter so the work of stars throughout their lives have produced little when compared to the whole, but just enough for our existence.

The prime mover for building the physical Universe is energy. When organisms (life) came along the flow of information, not energy, became the prime mover. A unit of life exists because "communication" of molecular information allows for survival in "entropy"[20] Universe. Once life existed Werner Loewenstein says: "Given its abundance, it is not surprising that evolution chose light as an energy source to drive living systems." Even the writers of the Bible

---

[20] The decay of any object, system and/or critter is referred to as entropy. An antique table is going through a process of entropy; a rusty nail; critters growing older; etc.

got it right when they wrote, before critters came along: "And then there was light."

A one cell creature, the Amoeba[21], has the capacity to process information that would fill more than 300,000 pages of a book. A human's body cells contain, collectively, enough information to fill a library greater than any that has ever existed.

Both living and non-living matter are made of the same kind of atoms, yet the living can perform coordinated functions (e.g. the cells in your heart muscles are the same type cells in your leg muscles yet the heart cells all have the same rhythmical beat) where the non-living cannot. Living cells do what non-living system cannot do, and that is extract information from their surroundings.

When a living creature is formed the building structure, cells, are given different job assignments. The aggregate chores are designed to make the whole body function. That is to do those things that give it the best chance to survive in its environment long enough to replicate the DNA matrix, assuring continuance of the specie. This is done by passing on information to and from each cell by a control center (for a human the communicator is the mind which resides in the cells of the brain). Entropy which in living entities is called aging, the flow of time, will win out after the next generation is produced. The process that results in "dust to dust" when we die is called entropy. Thus the disorder that our living body cells go through to become a "pile" of atoms is entropy.

---

[21] Loewenstein, Werner, "The Touch Stone of Life", Oxford Press, 1999

It's All About Time

The time flow on Earth has proven to be an ideal environment for complex life to evolve. "From Everest's peak to the floor of the Mariana Trench, creatures of one kind or another inhabit virtually every square inch of the planetary surface".[22] Earth's incubator has allowed microorganisms, such as bacteria and viruses[23], to become a diversity of life that is certainly "Earth like". Life that evolved on Earth can only survive elsewhere in an Earth like environment. However, microbes have proven that they can survive in space and there is some circumstantial evidence that the "seeds" of life planted on Earth did arrive from space.

A bacterium called "Deinococcus radiodarans" can live through radiation so intense the glass of a Pyrex beaker, in which they reside, will cook to a discolored fragile condition. These critters can with stand radiation about 1,000 times that which will kill Homo sapiens. A small number will survive three (3) million rads[24]. 1,000 rads will kill a human in one to two weeks. These guys are 'super bugs' and certainly are candidates for space travel.

When the first astronauts landed on the moon in 1969 they retrieved a piece of an unmanned moon vehicle of the "Surveyor" series that had been on

---

[22] Wilson, Edward O., "The Future of Life", Vintage Books, 2002

[23] A virus is a small infectious agent that can replicate only inside the living cells of organisms. Viruses infect all types of organisms, from animals and plants to bacteria and archaea. About 5,000 viruses have been described in detail, although there are millions of different types. Viruses are found in almost every ecosystem on Earth and are the most abundant type of biological entity. They appear to be the bridge between flora to fauna.

[24] A radiation unit i.e.-curie, roentgen, rad, and rem.

the moon surface for seven plus years. (The unmanned "Surveyor" instruments were sent to the moon to recon landing sites for the Apollo program.) Bacteria found on the retrieved part, exposed to the moon's environment, were still alive.

There was an experiment on board the space station that exposed microbes to space. The shuttle Challenger's crew was supposed to retrieve this experiment and return it to researchers on Earth. Challenger exploded on lift off and the microbe experiment remained exposed to a space environment for three plus years. Examination of the experiment found the bacteria had created a cocoon around some members of their colony. The bacteria inside the cocoon survived.

There is evidence that microbes can remain dormant indefinitely. About thirty million years ago a bee was killed by resin that over time turned into amber which served as its tomb until now. Within the bee's belly a microbiologist, Raul Cano, found more than two thousand species of bacteria and yeast that had survived those millions of years of entombment[25]. When environmental conditions in a pristine laboratory reached life support levels, these critters came alive!

Almost every global influenza outbreak involves microbes that are unique to that particular outbreak. The influenza pandemic of 1918-19 killed from 20 to 30 million people from every inhabited continent. Physicians were helpless as "The origin of this influenza variant is not precisely known". The accepted thought was that the virus originated in China where a rare genetic shift took

---

[25] Warshofsky, Fred, "Stealing Time", TV Books, 1999

place to change the virus. (Note: In 1918 China was essentially a closed society to the rest of the world thus the perfect escape goat for this theory.) This theory did not explain the facts that people worldwide were infected nearly simultaneously, not allowing time needed for human carriers to infect all the continents.

Immediately after Halley's Comet visited the Solar System in April 1910 Earth's orbit took the planet directly through the comet's plume. Water[26] latent dust and other debris entered Earth's upper atmosphere. Some researchers believe the influenza virus that caused the 1918 outbreak arrived via Halley's Comet. I believe that eventually the seeds of life will probably be found in the dust debris left by a passing comet, particularly the comets that originate in the Jupiter region. These comets are called "snowballs" and have ice/water with the same molecular structure as the water molecule on Earth. Comets from the Oort cloud, located in the outer reaches of the Solar System, have a different water molecule than that found on Earth. These comets are called "ice bergs". Whether they support life or not has not been theorized according to my research efforts to date.

The aforementioned evidence drives the search for life on Mars and elsewhere in space. The Viking Landers of the 1970s were designed to search for life on the surface of Mars. Four experiments were conducted involving collection of soil, applying nutrients and analyzing the waste that could have been generated by microbes. The first experiment used a radiation marker which gave

---

[26] Scientists have determined that Earth's water was delivered to the upper atmosphere by comets. A phenomenon that continues today.

a positive indication that something alive had processed the soil sample. Two experiments gave unexpected and puzzling results, while the fourth experiment gave a negative result. Later, these same four experiments were conducted on Earth using twin Viking instruments that ran the same experiments analyzing the waste from soil enriched with microbes, and all the devices failed to detect the life contained in those samples.

NASA has a meteorite that is definitely from Mars that contains a fossil that was left by a microbe. The meteorite was examined several years ago without a conclusion. However, since the exam new instrumentation has been developed and the sample is being revaluated. Conclusions are pending.

There are many clues that lead to the conclusion that the Universe is populated throughout with life in the form of microorganisms. The age of Earth almost certainly excludes time for life to originate. It is a logical assumption that the seeds of life originated elsewhere at the microcosm level. "Our DNA is simply too paltry to spell out the wiring diagram for the human brain[27]".

It's conceivable that Nature's purpose for a collection of atoms coming "alive" is to replicate the double helical structure we call DNA. These two chains are wound round each other and linked together by hydrogen (the oldest and most abundant element) bonds between specific complimentary bases to form a spiral ladder-shaped molecule. The molecule can encode vast amounts of data, save it, and pass it on through countless generations nearly error free. Because it

---

[27] Schwartz, Jeffery M., & Begley, Sharon, "The Mind & the Brain", HarperCollins, 2002

operates at the quantum level only a particular arrangement of atoms is possible[28]
.

All known life on earth, from bacteria to trees, including humans, are all descendants from a single ancestor and share the same kind of genetic material, i.e., DNA. Living cells use DNA to store genetic information and use the same code for turning this information into proteins and living organisms.

For a collection of atoms to perform the chores of life it would appear to take eons of time to synchronize. These crude groups of microbes existed in a slower time zone, thus in the beginning of microcosmic life there was no need for replication or subsistence. It was not until the "seeds of life" arrived on Earth that the need for renewal became parameters of necessity for continued existence. The primary consequences of the computational nature of the Universe are that it generates complex systems, such as life[29]". Consciousness may be an emergent phenomenon that does not exist at a lower level of complexity than mammals have evolved on Earth.

Everything that is a 'living' entity depends upon the flow of information between the living cells. The time clock within each creature appears to allow this communication to take place, thus the entity that defines us as being alive. The instance a living creature dies every cell that defined that creature ceases to function/live.

---

[28] Llyod, Seth, "Programming the Universe", Alfred A. Knofp, 2006
[29] Heudin, Jean-Claude, Editor, "Virtual Worlds", Perseus, 2004

It's All About Time

What some define as a "normal" human being may simply be an entity with an internal clock that's in sync with its surroundings! The life span of any earthy creature, that is to say the aging process, is defined by time and its ability to replicate.

Our concepts of life, from birth to death, are constrained by having only observed a single instance of evolutionary life. Our brains have not evolved to the point where original thought can occur. The brain is preprogrammed (learning) and relies upon techniques of synergy to accumulate knowledge. To be creative or an inventor are misnomers since we simply discover Nature's secrets. The current state-of-art in science cannot define why, what or how 'life' exists.

So the two questions that are most important to scientists, "How did life begin?" and "What is consciousness?" go unanswered partly because of evidence and mostly because of the fabric of our philosophical upbringing. Philosophy developed by the Greeks[30] in the 4th century B.C. suggested that the body had a "soul" and that soul was immortal. Our descendants will continue the quest for scientific knowledge about the basic nature of life's origins.

There appears to be an infinite number of unknowns that exist in the quantum domain that make up the ingredients that produce the whole of life. Scientists will never develop the wherewithal to discover the preponderance of factors of the quantum enigma that constitute life because the observer is part of that world. We are stuck in the grasp of an evolutionary process that continues to hone our senses for the purpose of survival and propagation, not quantum

[30] Guthrie, W.K.C., "The Greek Philosophers", Harper, 1960

observations. Our understanding of the quantum world will be inhibited by this lack of sophistication until extinction.

The time window for life to begin must have occurred shortly after the atom was formed. The question is where did microcrobe life form? We know that the microcrobe can live for millions of years. It's clear that the microbes that landed on Earth came from the comets that formed around Jupiter. But the question remains: Is this the only place that life formed as a product of comet formation, or did the microbes just hitch a ride on these comets because they provided a better environment than the comet from deep space, which had a different water molecule. Once we know the answer we will know if life exists throughout the Universe or only within our Solar System.

CHAPTER 9
THE MECHANICS OF EARTH'S CLIMATE ENVIRONS

"The more you understand it, the more you realize you don't have to answer the question of whether or not something is conscious in order to define consciousness". But, he said, it's important not to come up with a definition before we've understood all the elements that need to be encompassed in that definition, least we suffer what he calls "the heartbreak of premature definition," an intellectual dysfunction he believes many of today's consciousness scholars suffer from.[31]

"WEATHER, n. The climate of an hour. A permanent topic of conversation among persons whom it does not interest, but who have inherited the tendency to chatter about it from naked arboreal ancestors whom it keenly concerned. The setting up of official weather bureaus and their maintenance in mendacity prove that even governments are accessible to suasion by the rude forefathers of the jungle" — Ambrose Bierce

Global climate(s) science continues to be held hostage by a debate between the scientific and special interest communities. Attempts by special interests groups have driven scientists to *simplify* an extremely complicated global weather system. To date, no entity has produced a simulation model that has been calibrated with any definitive precision to the degree of simulating an accurate projection, other than statistically based forecasts, for any single local point on Earth. Current forecasting has produced an attitude amongst non-

---

[31] Moffett, Shannon, "The Three-Pound Enigma", Algonquin, 2006. Quote is from Doctor Daniel Dennett, Philosopher

scientists leadership causing the climatic debate to move away from credible scientific study to a "culture" that has changed the non-scientific answer to: "do you believe or not believe that "global warming" is happening?" Without any qualitative assessments!

Universities appear to teach science that fits the world that we have created, which is not the Natural world. In fact the "environmental movement" has shaped academia's curriculum in the natural sciences. Very few institutions, if any, teach science from the view point of "why" or "how". Few teach the analysis of ALL data that's possible to observe to arrive at a conclusion. The very opposite appears to be true. "We know the answer so let's design our observations to collect data to verify our fore gone conclusions. Emphasis is placed upon the effect and not the cause. The following are some examples.

DUST: There is always a thin layer of dust circling the Earth in the upper reaches of the atmosphere. Beautifully colored sunsets are clues, especially after a volcanic eruption. Thousands of tons of dust are blown off Earth's deserts each year. Dust is one of the regulators of Earth's temperatures. Water vapors, which cannot occur in the atmosphere without dust particles, are the main players in climate control. In fact every drop of precipitation, rain, snow or ice, must have a speck of dust around which to form. Water vapor that forms clouds must have particles to form. Without dust the humidity below 300% will not condense. Above 300% humidity water will condense on objects including humans.

Let's look at the record storms of 2011 that occurred in the south and Midwest of the USA. The thousands of wild fires burning in the southwestern part of the USA produced many tons of dust into the atmosphere. This dust contributed to the volatility of these storms. No weather model was used in forecasting nor did the meteorologists presenting these forecasts mention, in their interpretations of conditions the possible quantitative impact, that dust, as an input, from these wild fires had any effect upon what was happening weather wise.

Without dust from the Sahara Desert, the Atlantic hurricane season would not exist. The Caribbean Islands would consist of grey rock without dust from the Sahara, which produced the Islands' layers of top soil.

Dust may turn out to be the most important thermal climate regulator of all the culprits that science has assigned that roll.

Everything on Earth, alive or not, is in a constant state of entropy, thus turning into dust. The millions of tons released into the atmosphere each year have an enormous effect upon Earth climates, from precipitation to becoming filament to absorb and retain calories/heat.

Carbon dioxide: This gas is heavier than the air column therefore will not rise to the upper atmosphere period. A side bar: A half bottle of champagne is best stored in a refrigerator with the cork out. Research has proven that the carbon dioxide will not escape from the open bottle!

Misperception: Plastic bottles last longer in certified sanitary landfills than paper milk cartons: Everything in a properly designed land fill that was not once alive deteriorates at about the same rate over time. The chemical makeup of plastic may break down before paper. Landfills have not been around long enough to determine their preservation lineage. Methane from decomposing once-alive compounds is the only thing that escapes a properly designed landfill.

Man's activities affect global climate? Read on!

Paying attention to the climatic details by defining/creating a model simulation that can depict the interrelationships of all of Earth's regional climates has been lost in the debate. This lack of scientific "state-of-the-art" exercise to calibrate any model with ground truth data has caused non-technical folk to dominate the discussions. The continued gathering and analysis of meteorological knowledge seem to be futile exercises. Neither climatic knowledge nor its precise manipulation that is "cause and effect" driven gets added to a one-sided debate.

In an effort to present a definitive global system that is "cause and effect" driven, utilizing the known climatic components will begin to make some sense in the continuing debate. The reasons as to why global ice advancement is more likely to occur than any significant long term global warm up are also presented. Included is the scientific rationale as to why **Antarctica is Earth's thermostat**.

Anyone can observe the vapors from dry ice (CO_2), at room temperature, as they fill up a container and fall over the edge like a water fall. Since CO_2 (390 ppm[32]) resides in the upper atmosphere, many researchers have searched for the mechanism which could lift CO_2, a gas heavier than the atmospheric gases, to the upper reaches of the atmosphere. One solution comes from the observation of data collected from the "Deep Impact" spacecraft[33] as it flew within 435 miles of

---

[32] Number of parts per million parts

It's All About Time

Comet Hartley 2 on November 4, 2010. Observations from this close encounter showed chunks of water ice being expelled from the comet in jets of CO_2. This is the first time CO_2 jets have been observed coming from a comet. Why the NASA environmental scientists have not recognized this source of CO_2 for the upper atmosphere baffles me.

My theory is that most of the breathable oxygen and all of Earth's carbon came from "snowball" comet in the form of CO_2. Because the molecule CO_2 is heavier than the atmosphere gases (atomic weight 0f 44 vs 34) the CO_2 slowly sank toward the surface. The flora on Earth processed[34] the CO_2 retaining the carbon and releasing the O_2.

From about 200 million years ago (Triassic geologic period) until about 65 millions years (end of the Jurassic period) the Earth must have been bombarded with "snow ball" comets which fed CO_2 to the flora that produced huge amounts of O_2 which made the fauna (critters) robust. Nature using this process produced all the fossil fuels that we enjoy today.

After 65 million years the comet bombardment moderated thus cutting off the fuel supply for the flora, which could not produce enough O_2 to support critters as large as the dinosaurs.

---

[33] The Epoxi mission (www.nasa.gov/mission pages/epoxi/)
[34] Photosynthesis is a chemical process that converts carbon dioxide into organic compounds, especially sugars, using the energy from sunlight. Photosynthesis occurs in plants, algae, and many species of bacteria.

It's All About Time

In retrospect the "snow ball" comet brought water, CO_2 and microbes to an early Earth and this process continues today.

Research shows that the breakup of another "snowball[35]" comet, Comet LINEAR[36], was likely made up of water with the same isotopic composition as water found here on Earth. The finding supports a controversial idea that comet impacts billions of years ago could have provided most of the water in Earth's oceans. "The smaller comets from Jupiter's region impacted Earth relatively gently, shattering high in the atmosphere and delivering most of their organic molecules intact"[37].

There are two (2) questions plus more to follow that need to be defined before we assess how the CO_2 got to the upper atmosphere: (1) what part of the infamous "Atmospheric Carbon Dioxide" graph depicting the measurement at Mauna Loa, HI, is caused from entropy resulting from activities on Earth and/or comets? And, (2) could not the heavier than air CO_2 been deposited by a comet and settled through the atmosphere to the lower reaches? (The latter being the most likely scenario.)

While flying for the military, I have, on occasion, observed a clear path with unlimited visibility created from the precipitation from a thunder storm as it passed through a hazy air layer in the lower atmosphere that had a visibility of

---

[35] Comets from deep space formed in an extremely cold region thus are called "ice bergs" while comets formed in the Jupiter region, a more moderate cold, are referred to as "snowballs".
[36] August 2000
[37] *The Science and Technology Directorate at NASA's* <u>*Marshall Space Flight Center*</u>*, May 18, 2001*

less than one mile. Precipitation appears to be the cleansing agent for the lower air column. Thus, any gas produced on Earth that has a density higher than the atmosphere will not go beyond the influence of the **hydrologic cycle**[38] which produces all of Earth's precipitation.

What would happen if Earth's surface temperature continues to rise a few degrees in the next couple centuries? What would be the effect upon its inhabitants? When would the Antarctica thermostat stop this rise in temperature?

Earth's climate is an aggregate of regional climates working as a global system driven by orographics[39], plate tectonics, thermodynamics, astrophysics and kinetics. This is to say mountains, moving continents, deserts, plains, hot or cold liquid (air and water) interfaces, Sun, and moon, all provide a force that results in the coordinated movement en masse.

The oceans are the "heat sinks" that keep the global temperatures moderate. Ocean currents, called conveyors or streams, carry the equatorial heat to the Polar Regions. The conveyors are guided by the continents and the Earth's rotation. If a conveyor is stopped, the applicable tropics will heat up and the cold

---

[38] Water exists on earth as a solid (ice), liquid or gas (water vapor). Oceans, rivers, clouds, and rain, all of which contain water, are in a frequent state of change (surface water evaporates, cloud water precipitates, rainfall infiltrates the ground, etc.). However, the total amount of the earth's water does not change. The circulation and conservation of earth's water is called the "**hydrologic cycle**".

[39] Highlands and mountain ranges

latitudes will become colder. If a conveyor speeds up the heat distribution will lessen, thus temperatures, over time, will also fall in the Polar Regions.

Air is less dense therefore has minimal effect upon Earth's temperatures as compared to an equal volume of water. For example, you can spend more than 20 minutes inside a dry sauna at temperatures of 65°C (190° F), but you can't hold your hand under a faucet of hot water at 36°C (110°F) for more than a few seconds.

The globe is covered with distinct weather systems that overlap, interact and thus cover the entire globe. The Gulf Stream is one example of a weather system that affects a regional climate.

The regional climates consist of a series of distinct local climates. An example of a local climate is the Los Angeles Basin. This "bowl", created by the mountains, under certain weather conditions, can cause smog to form because of exhaust from a high concentration of automobiles. One weather system moving through an area can refresh/scrub the local climate.

Before several continuous scenarios of assumed rising temperatures can be played out, the major components such as ice, water, solar, celestial, internal radiation, state-of-the-art modeling, ocean conveyors, continent locations, etc., should be collectively described as they relate to Earth's climates. At best with today's technology, we can vaguely comprehend the sophistication of how an infinite number of micro climates could result, via the "butterfly effect", into multiple environments over time.

It's All About Time

Before we can comprehend what ingredients constitute a global "climate", we must have an awareness of the magnitude, diversity and difficulty that this understanding may pose. We know that knowledge of a local climate cannot readily be extrapolated to a global one without the analysis of a millennium of data. To enter this data in today's modeling would be nearly impossible because of the modeling techniques. If it were done, the results would simply be unreliable! One example is the aggregate composition of models used to create a 'statistical fan' to describe the future track of a hurricane's path. Thousands of 'real time' data observations don't change the forecast from being parts scientific and experience.

Short sightedness that ignores the magnitude of a global translation that requires extrapolation of local visual evidence (i.e.-retreating glaciers) as a global condition is like picking a single cell from your body to evaluate your body's health! We bias the results by "knowing" the answer before we collect and analyze the data to arrive at a conclusion.

Earth and the moon rotate around each other, as if tied together with a string, as they travel around the Sun in an orbit that has never been fixed. Earth is a gyroscopic sphere flying through space at more than 69,361[40] miles per hour. It tilts, wobbles, precesses and is affected by everybody in the Solar System and

---

[40] Spin and orbit speed = 69,361 mph; Earth is moving toward Lambda Herculis at 43,200 mph; Earth's motion perpendicular to Galactic Plane = 15,624 mph; The Galactic spin rate =446,400 mph; and if you left the Milky Way Galaxy add another 1,339,200 mph, the speed the Milky Way is moving through the Universe.

beyond. Since Earth was formed some 4.66 billion years ago, it has never experienced any moment like any other! For example, after the winter solstice "the position of the perihelion (nearest the Sun) shifts steadily and makes a complete circuit of orbit in 21,310 years. The actual amount of the tilt changes very slightly, growing a tiny bit more, then a tiny bit less, and in slow oscillation. All of these changes have a small effect upon Earth's average temperature, not great, but enough at certain times to pull the trigger for either the advance of glaciers or their retreat.[41]"

900 million years ago Earth's day was 18 hours[42] long and a year[43] was 481 days duration. The effect of the moon has caused the changes that are today's conditions. Future days will become longer and the years will become shorter. Future building of wind generators will exacerbate the slowing of Earth's rotation since this is the energy used by the generators to produce electricity.

Earth's surface is a water domain. Water covers about 71% of the surface yet it constitutes only about 1/4200th of Earth's total mass. Oceans hold about 97.2% of Earth's water and are the source for fresh water to the tune of 80,000 cubic miles evaporated each year that fall in the form of rain or snow. Stored underground are some 200,000 cubic miles of water, mostly fresh, with an additional 30,000 cubic miles stored in lakes and rivers.

---

[41] Asimov, Isaac, "New Guide to Science", Basic Books, 1984
[42] The time it took for one complete rotation.
[43] The time it took for one orbit around the Sun.

It's All About Time

Water, in a solid state, covers about 10% of the Earth's surface which is roughly the size of the North American continent. The Antarctic ice sheet contains about 91% of the total ice on Earth. Greenland has about 8% of Earth's ice while the mountain glaciers and Arctic cap account for less than 1% of the total.

The following scenarios are based upon an assumption of continued rise in Earth's temperatures to the point where cooling will obviously begin in most regions.

Arctic's ice cap floats. Its position at any given moment is at the whim of the ocean conveyors, continental boundaries and prevailing winds. The melting of the Arctic ice will be from calories given up by the ocean and not the atmosphere. Since this ice total is less than one per cent of Earth's ice the effects, if melted, are local only. The ebb and flow of these ice packs are a result of a regional climate, not global.

If ALL the ice from the Arctic and glaciers melted at the end of the summer the effect upon sea level would be negligible, nearly impossible to ascertain since the oceans of the world are subservient to wind fetches and tidal gyrations from a few feet to many tens of feet. However, there is evidence that this fresh water melt, augmented by the Greenland melt, did overlay the cold Arctic salt waters to the point that it shut down the Gulf Stream (GS) for 10 days in 2004. This indicates that this regional climate is very fragile from a thermal observation.

It's All About Time

No scientist knows what caused the GS to stop flowing in 2004. According to the Scientists at Woods Hole, the stoppage event was described as "the most abrupt change in the whole (climate) record".

What would happen if "a significant amount of Greenland's ice cap melted"?

The latest climate models predict that the GS will slow down as global warming increases. However, measurements by NASA[44] of the Atlantic Meridional Overturning Circulation show no significant slowing over the last 15 years; in fact the data suggest the circulation may have sped up by as much as 20% in the recent past.

The GS is the conveyor that keeps England and northern Europe from having a regional climate similar to the climate in Canada above the 45th latitude. Warm surface water flows from the tropics northward into the North Atlantic as one of the currents that make up the Atlantic overturning circulation system. Within the oceans surrounding Greenland the GS cools and sinks to great depths as it changes direction. What was once warm surface water heading north becomes cold deep water heading generally south.

The GS starts from the Equatorial Current from the African coast, moving east to west under the influence of the trade winds in the tropical North Atlantic. The South American continent deflects the current northward causing it to meander among the Caribbean Islands. The Equatorial Current circles the Gulf of

---

[44] Jet Propulsion Laboratory, Pasadena, CA Press release dated March 25, 2010

It's All About Time

Mexico in a clockwise fashion, exiting through the straits between Florida and Cuba. Then the Stream joins the Antilles Current, officially forming the GS.

The GS is about 90 kilometers wide and flows at two meters per second at about 60 degrees latitude. The GS flows at about 80 million cubic meters per second, which exceeds the volume of ALL rivers in the world. The volume of the GS is 3500 times larger than the Mississippi River's discharge into the Gulf of Mexico.

The large volume of warm water moved by the GS toward the colder North East Atlantic reaches near Latitude 40-42 degrees north before it's deflected southward. The GS loses heat energy by melting the ice floes, as well as calories loss as the cold fresh water, from the glaciers, all of which overlay the GS cooling it to a density[45] of the surrounding salt water. The result is the GS loses its identity and becomes part of the North Atlantic Ocean.

If melt from the Greenland ice pack increases, there will be an increase of fresh cold water with a density of 1.000 overlaying the ocean of cold saltwater with a density of 1.030. The boundary[46] integrity of any two liquids of different densities is very rigid.

Greenland's ice cap volume is about 2.85 million cubic kilometers. If all the ice melted, the mean elevation of the world's oceans would be increased by about 23.6 meters. But, more realistically, for each 100 meters of ice melt equivalent to the Island size, the oceans would rise about 19.5 inches.

---

[45] Density of the cold salt water = 1.030+; Ice melt density = 1.000
[46] The boundary is called a thermocline.

It's All About Time

Since Greenland on average is warmer than Antarctica, an increase in local temperatures could produce melting here first. If the temperatures on Greenland continue to rise, the snowfall will increase on the ice cap. This will increase the ice cap volume and provide more ice for glacier calving into the North Atlantic. This is why it's uncertain if the ice sheets on Greenland and Antarctica are growing or shrinking. Antarctica is so cold that surface melting will not occur, but Greenland is a different story by 50+ degrees Fahrenheit.

Once the GS is cut off, the regional climates of northern Europe will no longer be the recipient of the tropical heat energy. With time, their climate will emulate that of Canada above the 45° latitude. The Polar ice cap will grow to include the North Sea and will attach itself to the continent. Ocean currents, which are the conveyors of surface energy around the globe will be modified, energy wise, to the point that the Arctic ice cap will, over time, expand. As the ice cap grows heat from the Sun will be deflected and the ice pack will continue to enlarge. Greenland's seasonal temperatures will start to decrease, the ice pack will begin to grow and the warm-cold cycle will continue as it has historically.

Northern Europe and the North Sea oil platforms will become uninhabitable over time. The increase in ice coverage will reflect the Sun's energies during the summer, and the Earth will begin a cool cycle. The increase in ice coverage will take water from the oceans thus they will recede.

What regional climate changes will result from a continued warming trend, say an extreme four to six degrees Fahrenheit, for the next two centuries?

- Here's my list:

It's All About Time

- *Antarctica's annual precipitation would nearly double over time taking moisture from the oceans.*
- *New York would become a river city.*
- *The corn and wheat belts of Iowa and Nebraska would slowly move northward to the plains of Alberta and Saskatchewan.*
- *All continents would gain land mass from the oceans receding exposing continental shelf as the ocean would fall nearly 450 feet in a couple centuries. The Florida Keys would become a peninsula.*
- *Greenland's ice pack would begin to grow as the glacier movement slows and calving of ice also slows.*
- *Ice sheet covering the South Pole would expand at an alarming rate.*
- *Southern oceans cluttered with ice floes would cause water temperatures to drop significantly.*
- *Antarctic's research station would be abandoned because of ice movement caused by an increase in precipitation.*
- *Annual precipitation in the USA's upper Midwest would double to 20+/- inches. This increase in runoff would, over time, cause the Mississippi River's annual flow to gradually increase to the point that diversion around New Orleans would become necessary only during major flood events. As the river channel eroded because of the receding Gulf levels, New Orleans would eventually be above sea level.*

It's All About Time

- *Oceans would cool three to five degrees C. This would cause the air temperature to cool. Local areas would experience ice ages which would shorten the summer season.   The polar bears would return to the ice.*
- *If the precipitation in Antarctica doubles, the moisture must come from the oceans. Since the continent is so large an additional inch over this area for two hundred years constitutes a large volume of water.*
- *The Southwest US is nearly a desert now so a very little rise in temperature would cause the local climate to continue in that direction.   Since the weather pattern in the US travels from the Southwest toward the northeast, there would be more dust available to create an increase in precipitation down range. Thus, weather systems in the US may become more volatile.*
- *Population in Egypt would expand beyond the Nile Valley as the desert blooms.*
- *The Atlantic hurricane season would nearly cease as the Sahara desert blooms.*
- *If the Sahara Desert blooms, there would be less dust available to form clouds and/ or rain drops. If a hurricane does not have clouds and/or rain to dissipate its energy, will it self destruct?*

Increasing regional temperatures over time would cause the global temperature to cycle to a cool down. There are several "safeguard" mechanisms in place to prevent a warm up, but none to prevent an ice age. During Earth's history, Antarctica has proven to be its thermostat.   Earth has never been overheated since the initial cool down and as long as the thermostat is in place, it will not be.

Before the study of plate tectonics became a science, some people as early as 1596[47] believed that the arrangement of continents appeared to be puzzle pieces that could have fit together to form a super continent in years past. Currently the "theory of continental drift" suggests that some 225 to 260 million years ago all seven continents were together forming a super continent called "Pangaea"[48]. For more than 225 million years, the Antarctic continent has remained near Earth's "bottom" while the other continents, North and South America, Europe, Asia, Africa and Oceania (Australia), have drifted[49] to their present positions. Evidence from oceanic ridges surrounding Antarctica indicates that the super continent began to break up about 150 million years ago. Fossils, soil, rock, modeling and other evidence support these conclusions.

---

[47] In Dutch map maker Abraham Ortelius' work "Thesaurus Geographicus" he suggested that the Americas were "torn away from Europe and Africa by earthquakes and floods". It was not until 1912 when the idea of moving continents was seriously considered as a scientific theory called "Continental Drift", introduced by a German meteorologist Alfred Lothar Wegener.

[48] USGS (U.

[49] Today these continents continue to drift at the about the same speed that our finger nails grow. (US Geological Survey USGS) The "theory of plate tectonics" states that **Earth's** outer most layer consist of a dozen or more plates that are moving relative to one another as they float on hotter, more fluid material.

It's All About Time

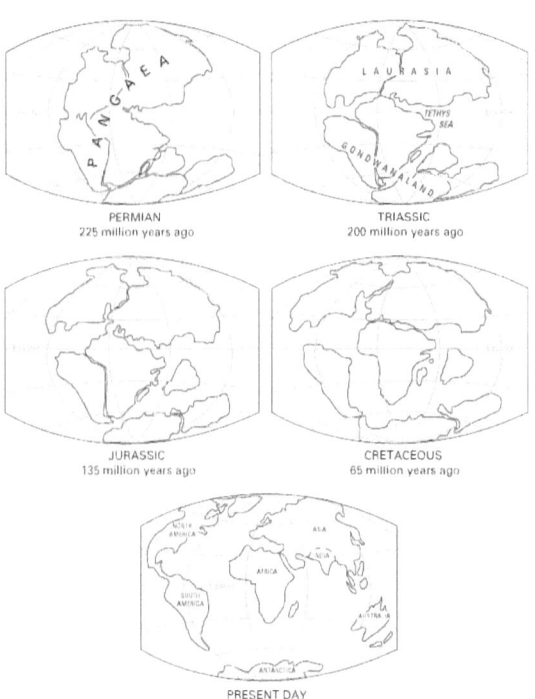

Plate tectonic history (from Bing.com)

The Antarctic, on average, is the coldest, driest, and windiest continent, and has the highest average <u>elevation</u> of all the continents. Antarctica[50] is the

It's All About Time

fifth largest continent by land mass, and the third largest, if the area of the ice cap is measured, assuming that Greenland is part of the North American continent which is the second largest. The summer temperatures rise to about minus 30°C to Earth's lowest yearly air temperature of about minus 80°C. The relative humidity is 0.04 % (Mars' is 0.03%) making the continent the driest place on the globe. Annual precipitation averages just above an inch, mostly from ice fog over the interior regions.

Earth's rotation creates a gyroscopic effect that has appeared to lock Antarctica in its current position at the "bottom" of the globe. Forces from a gyroscope react along a three dimensional axis, separated by 90 degrees; e.g. spin, output and input. Earth's spin axis is the global center. The Sun provides the input, and the output axis is 90° from the input. Thus it affects the South and North Poles. The continent lies 90° from the centroid of a gyroscope, Earth, and its gravity "string" from the Sun. The effect of the forces created by the continental drift appears to be inhibited by this gyroscopic effect. *Or is it something simple like the Antarctic continent continues to float on the apex of a globe that is spinning!*

During the Triassic and Jurassic periods, 200 to 150 million years ago respectively, when most of the ingredients for fossil fuels were laid down, the dinosaurs grew to be very robust because of an abundance of O_2 in the atmosphere. The Earth warmed up a bit, as its thermostat worked perfectly. Had

---

[50] Winkel-tripel-projection.jpg

Homo sapiens been around at that time, they too would have been more robust because of the abundance of oxygen.

Therefore, Antarctica has been in place for more than 225 million years and will remain as Earth's thermostat for many more millions of years, preventing future significant global warm-ups.

The following highlights one theme from my previous list that deserves a further note:

In the early 1600's when Europeans first arrived upon these shores, the Appalachian Mountain range was covered with mature trees (virgin forest). The Appalachians run from Maine to Georgia. Most of the "virgin" timber has been harvested while some second growth forest has been preserved as National Forest. Hardwood forest in these forests will take from 150 to 500 years to develop old-growth characteristics in one or two generations of trees.

Cathedral State Park in West Virginia consists of only 133 acres of over 170 species of vascular flora that include 30 tree species of which 17 are broad leaf, nine species of ferns, three of club moss and over 50 species of wild flowers. The flora is as it was in the early 1600s, a microcosm of the historical Appalachian forest. The trees include virgin hemlock up to 90 feet high and 21 feet circumference form cloisters in the park. A soft wood tree of this size has a carbon content of 1265 Kg[51] and a CO_2 equivalent of 4638 Kg. A virgin hardwood with an average diameter of 10 feet would have about 5542 kg of carbon and

---

[51] One pound = 0.453592 kilograms.

It's All About Time

20,322 kg of CO_2 equivalent locked up. The large oak tree next to my door has about 15,000 kg of oxygen locked up that, if released, would make Homo sapiens more robust.

Did you ever wonder how the dinosaurs evolved to be so large? Or why so much of today's fossil fuels originated during this same period? The flora and fauna grew so lush because carbon dioxide was available in the atmosphere. During this time the growth rate excelled causing huge critters to evolve because oxygen was in the air!

Increasing local temperatures over time will cause the global temperature to cycle to a cool down. During Earth's history there have been safeguards against warming, Antarctica being the thermostat. Earth has never been overheated and as long as the thermostat is in place it will not be.

We as a scientific society have learned to construct devices that can perform measurements with extreme relative precision. Yet with today's technological knowledge and computational tools, we can only define weather in terms of probability or chaos theories. Our scientific advancements have been exceptional during the last 400 years, but we still have only begun to understand. We still have a few millennia before we can begin to create a model that can, with precision, emulate the "butterfly effect", therefore allowing us to make a precise forecast for a specific point. Once this milestone is reached, then we will have begun to advance. To date Homo sapiens's effect upon the globe can only be measured locally and NOT regionally or globally. Our arguments should be directed toward the sophistication of the science that defines the climate of our environment.

The world that we have created to live in, its causes and effect consequences as described in chapter 8, is more philosophy driven (special interest) than Nature driven. One group of "special interests" verses another group with a conflicting interpretation has created a near "virtual" reality of how Nature works. Unfortunately the debate is driven by non-scientists leaders who have determined the objectives and "cherry" pick the data to support the desired result. How does this all happen?

The virtual world that we have created makes no one accountable for the group's actions. Sometimes the solution is worse than the cure. For example take the wild fires that swamped the southwest during 2011. Years ago timbering was halted in the National Forests because an owl's habitat would be affected. No degree or significant was place upon a scientific analysis to quantified the degree of harm that selective timbering would accomplish. Emphasis was placed upon the group that would profit from the endeavor. This emphasis reflected the totality of the issue. But the lack of harvesting selected trees over time causes a buildup of fuel on the forest floor. Thus the result is a fire that difficult to control, and the ultimate goal of protecting the owls is destroyed. There are many instances of similar disasters created by a "cure".

We are who we are by accident of birth and we quickly bond with our "tribe". This is human nature. The success of sports teams, armies and "causes" depends upon the bonding of the individuals. "They jelled", as a collective whole.

These "team" members are considered to be the individuals that perform "inside" the envelope. Each represents what is expected in this world that we have created. Only the individuals that survive on the outer boundaries of the envelope, or even outside the envelope (beyond societal acceptances), make a contribution or advance the science of "being". These individuals probe the secrets of how Nature works. Thus they are not bound by artificial rules of the "world" that society has created.

This is my objective; to arrive at a probable effect based on the knowledge that we have accumulated to date with no preconceived notions or results!

It's All About Time

CHAPTER 10
EPILOGUE

"The questions that define physics are "how" and "why"; It is answers to the "why" questions that take us to limits of knowledge". Neville McMorris[52]

The lineage of today's science probably began to mature in the 16th century as a small crack began to open between science and philosophy. The apron strings of religion were beginning to loosen. A period that some historians believe began the awakening of present day science.

Kuhn[53]describes the extent of the "philosophical" dispute about the nature of science between the realist Max Planck (1858-1947) and the instrumentalist Ernst Mach (1838-1916). He considers the outcome of their dispute gradually to be taken for granted by philosophers of science, yet culminating in the Orewellian sensibility that he says is so integral to science's sense of its own history. (Fullers book is worth a study of the philosophy difference between what Nature represents and realities that we have created.)

Throughout the history of science, there has not been a cordial relationship with religion. Until about the 19th century, the "Church" has been the constant constraint authority on the free development and communication of scientific principles. However, over the centuries philosophy has not advanced in changing its doctrine to reflect the scientific discoveries that have been made. As a matter

---

[52] McMorris, Neville, "The Nature of Science", Fairleigh Dickinson Universe Press, 1989
[53] Kuhn, Thomas; Fuller, Steve, "A Philosophical History for Our Times", University of Chicago Press, 2000

73

of fact, some prominent scientists, particularly physicists, believe that philosophy is near its intellectual end because of a lack of conforming to the knowledge that the world has gained about how Nature works. Martin Luther[54] is credited with the statement "that the biggest danger to religion was men who could reason".

A well documented conflict between religion and science involved Galileo Galilei (1564–1642), an astronomer, philosopher and physicist. He built the first telescope. After pointing it toward the stars, he soon discovered that Earth was not the center of the Solar System. The Sun was the center!   Since this was against "Church" doctrine, he was put under house arrest for the remainder of his life. Had he not known the Pope nor was a famous scientists of that time, he most likely would have been burned at the stake.

By the second century B.C. the Greeks knew that the Sun was the center of the Universe. However, the church kept a lid on that knowledge for 1700 years.

Some philosophy teachers believe that the United States was created as a religious nation. There is a Treaty with Tripoli, dated 1796, that states that the U.S. was NOT founded as a religious nation. John Adams, an atheist, signed the treaty. Thomas Jefferson, who wrote the Declaration of Independence, was also an atheist; Benjamin Franklin was an agnostic, and George Washington never attended church. What these patriots believed in was that every individual had the freedom and liberty to believe in whatever they chose.

---

[54] Luther, Martin (10 November 1483-18 February 1546) A German Priest, professor of theology and iconic figure of the Protestant Reformation.

It's All About Time

The motto "In God We Trust" was adopted in the mid 1950s, along with the phrase added to the Pledge of Allegiance "under God". Today polls show that more people do not practice religion than do in the U.S., so maybe Stephen Hawkins (English scientist, physicist, and mathematician) is right when he writes in his new book, "The Grand Design", that philosophy has outlived its usefulness.

Today there are conflicts concerning what is taught in public secondary schools that involve the separation of church and state. This separation issue is very vague amongst most politicians and political parties and their philosophies. (e.g.-the religious right, etc)

I am not a student of philosophy, but I have studied the Greeks from about the sixth century B.C. They are the originator of the philosophy that our virtual world was built upon. This virtual world dominated society for about 18 centuries. Beginning with Galileo's questioning the science of the natural world science has gradually exposed the curtain that hides the "Wizard of Oz", a.k.a. the Church.

Also, I do not wish this book to become a referendum on philosophy other than an honorable mention of its place in this world that we have created. Therefore I'll end this thought with a quote from Protagoras, 4th century B.C.: "Concerning the gods, I have no means of knowing whether they exist or not, nor of what form they are; for there are many obstacles to such knowledge, including the obscurity of the subject and the shortness of human life".

Scientists have also been a constraint to the advancement of pure science. Besides the claim for the success of "cold fusion" there are many questions about research results that are well documented[55]. One of the best published works that "uncover the intellectual rivalries, petty jealousies, and faulty science behind one of the most famous experiment in the history of evolutionary biology" is Judith Hooper's "Of Moths and Men", Norton, 2002. The entire research was staged.

Another colossal example was the National Geographic Society's (NGS) efforts in trying to suppress the news from Kitty Hawk, NC, that the Wright brothers had successfully performed power flight, while The NGS's sponsored flight attempt with Langley in Virginia had continued to fail. There have been other reservations expressed concerning the NGS's sponsored expedition to the Poles.

In an article, "World Without Ice", published in the National Geographic Magazine, dated October 2011, the researchers present an analysis of the climate during the end of the Paleocene geologic period (56.1 million years ago) and into the Eocene period based upon analysis of core samples taken from deep in the Earth's crust. The conclusion of the researchers was that carbon[56] from an unknown carbon dioxide source caused this climate condition. This part of the

---

[55] Hogan, James P., "Kicking the Sacred Cow", Simon & Schuester, 2004[55] Hogan, James P., "Kicking the Sacred Cow", Simon & Schuester, 2004

    Kruszelnicki, Karl, "Great Myths-Conceptions", A. McMeel, 2006
    Smolin, Lee, "The Trouble with Physics", !st Mariner Books, 2006
[56] NOTE: It appears that all of earth's carbon is found in its crust, another clue that the element's source was from above.

conclusion is probably correct because the source of the carbon dioxide was from comets that visited Earth from the Jupiter region. However, this climate condition can NOT be applied to the argument that the entropy from burning fossil fuels, (e.g., coal fired power plants) in the 21st century, will, in time, produce a similar climatic result. This leap of science is unsustainable since the carbon dioxide in the upper reaches of the air column has always arrived from space and NOT from any activity on Earth's surface.

There are many more documented stories about questionable scientific results that have occurred throughout the history of science.

Science is getting close to the obsolescent limits of the current mathematical language. A new language must be discovered before we can begin to comprehend aspects of the quantum enigma in terms of understanding how to repeat its concepts. Our mathematical systems are not intricately flexible nor precise enough nor free of "constants" (fudge factors) to affect dependable results from similitude exercises that can be modeled with "acceptable" accuracy. More scientific energies must be committed to this effort than are currently being applied to creating more eloquent cosmological geometries! There is evidence that the laws of physics change as the Universe evolves[57].

One major example of the stagnation in science is the "superstring theory, model or hunch" that has not moved from square one in a quarter century. With no prospects beyond eloquences in sight, the subject still drives many scientific cultures.

---

[57] Brooks, Michael, "13 Things That Don't Make Sense", Doubleday, 2008

It's All About Time

In early November 2008, I was getting my annual eye exam with a doctor that I have known for some time. He was a military flight surgeon, and I flew military aircraft so we have always had something to talk about during each visit. Our conversations continue as if no time had transpired in between. During this visit we were talking about the "dark" side of science which prompted him to show me three different vials of eye drops. "Research" has show that each one is best for glaucoma. Each research endeavor was conducted by scientists that had PhDs in the field. The three efforts were financed by the respective manufacturer of the eye drops.

Science has had a dark side since its birth. The pressure on many scientists has biased the outcome of their work, either by constraints placed upon them by those who control the resources, such as time, funds and/or institutions. Unfortunately too much research is conducted for the purpose of proving a preconceived notion or jerry rigged to prove an outcome from a controlled experiment; that is to say that predetermined data are collected and models are constructed to prove a preconceived conclusion. If a researcher's conclusions do not readily agree with the scientific community's accepted positions, it is almost never cited in their papers. The" incestuous" scientific world tries to remain closed to those folk who dare not challenge their views. Fortunately, significant advancements have been brought about by pioneers who dare to challenge the "community". These are the scientists and engineers who chose to imagine outside the conventional scientific envelope! They did not collect data to help generate a cause to satisfy a "preconceived" scientific notion, but they allowed the analysis, of all the available data, to lead them to a conclusion that has advanced the knowledge of science.

The acquisition of grant funds has become a cottage industry. The competition facing many grant requesters leads to the submission of grant applications containing fraudulent or questionable data. My experience with granting funds has lead me to develop an understanding of the problem(s) to be "solved" and the capabilities of the recipient to solve those problems. Sometimes by accident someone discovers useful new science that makes the process worthwhile. Most academic grants are used to educate students toward advanced degrees, thus the payout will accumulate over the life of their careers.

I know of no "pure" research programs being conducted. Most research can be classified as applied research, to solve a known problem

One last point about nonexistent research results concerning DDT- the "bug' killer. There was never any research that suggested that DDT was harmful to man and/or fowl. In Africa with the use of DDT the death rate from malaria and river blindness was nearly zeroed out. After special interest groups successful lobbying of the Federal Government to ban DDT using fraudulent data (the eagles along the Potomac were having trouble with egg shells for more than a decade before DDT hit the market), the death rates had again climbed into the millions per year. Are special interest folk murderers?

JUST 'CAUSE' SOMEONE SAYS IT DOESN'T MAKE IT SO! But it does add to your body of knowledge. In the early 1800's a peasant asked an engineer: "How does a steam engine work?" The engineer drew several charts and plans to show the peasant how the fuel converted the water into steam and how the steam powered the steam engine. The peasant said: "I understand completely but where is the horse?"

Well I understand my passions completely, but I have no idea where the horse is!

Any unification of Universal structure will/must enter through the portals of time. The "Rosetta Stone" of the "unity theory[58]" must be the time key.

"If we end up with a coherent and consistent unified theory of the Universe, involving extremely complicated mathematics, do we believe that this represents "reality"? Do we believe that the laws of nature are laid down using the elaborate algebraic machinery that is now merging in string theory? Or is it possible that nature's laws are much deeper, simple yet subtle, and that the mathematical description we use is simply the best we can do with the tools we have? In other words, perhaps we have not yet found the right language or framework to see the ultimate simplicity of nature." Michael Atiyah (one of the greatest mathematicians of the second half of the 20th century.)[59]

The works of Nature are absolute. There are no constants, chaos, probabilities and static states. The whole and all of its components must follow the same 'laws'. Time allows Nature to progress in a predetermined way throughout its components microcosm and macrocosm. Natural selection through evolution has given Homo sapiens the ability to focus our mind's eye in

---

[58] Theory of everything!
[59] Woit, Peter, "Not Even Wrong", Basic Books, 2006

whatever direction we freely choose. Our world will, in time, die as the Sun dies, but nature will continue!

Everything that I write about means nothing to most of us because these theories do not affect our daily lives. We have absolutely no control over any aspect of Nature because we exist as part of it. Our quest will always be to gain knowledge of how Nature works. This quest began when the Greeks began to look into the scientica of how Nature works. They started questioning the "power" of their Gods. As a result the Greek Gods slowly became part of their mythological history, thus, a non-entity in their daily life chores. In time 'scientica' became 'science'.

Today the accumulative knowledge curve of science is just beginning to move upward. As we discover Nature's secrets, we continue to add to the aggregate body of scientific knowledge that will greet our descendents. So the future is real!

Nature is what it is, not otherwise!

It's All About Time

SUGGESTED READING

1. Abrams, Nancy Ellen; Primack, Joel R., "The View from the Center of the Universe", Riverhead, 2006
2. Adams, Fred & Laughlin, Greg, "Five Ages of the Universe", Simon & Sch., 1999
3. *Asimov, Isaac, "Asimov's New Guide to Science", Basic Books, Inc., New York 1984*
4. Asimov, Isaac, "Atom- Journey Across the Sub Atomic Cosmos", Plume, 1992
5. Auyang, Sunny Y., "How is Quantum Field Theory Possible?", Oxford Univ., 1995
6. Barbour, Ian C., "When Science Meets Religion", Harper, San Fran., 2000
7. Barratt, Krome, "Logic & Design in Art, Sciences & Mathematics", Design Books, NY 1980
8. *Barrow, John D., "The Book of Nothing", Pantheon Books, New York 2000*
9. Bartusiak, Marcia, "Einstein's Unfinished Symphony", Henry, 2000
10. Behe, Michael J., "The Edge of Evolution", Free Press, 2007
11. Benacciho, Professor L, "The Great Atlas of the Universe", D&C, 2007
12. Berry, Adrian, "Galileo & the Dolphins", Wiley, 1996
13. Bloom, Howard, "Global Brain", Wiley & Sons, 2000
14. Boslough, John; Mather, John, "The Very First Light", BasicBooks, 1996
15. Boss, Alan, "Looking for Earths", John Wiley, 1998
16. Bova, Ben, "Faint Echoes, Distant Stars", HarperCollins, 2004
17. Brockelman, Paul, "Cosmology and Creation", Oxford University Press, 1999
18. Brooks, Michael, "13 Things that don't make sense", Doubleday, 2008
19. Bruce, Colin, "Schrodinger's Rabbits", Joseph Henry Press, Washington, D. C., 2004

20. Bryson, Bill, "A Short History of Nearly Everything", Broadway Books, Random House, 2003
21. Buchanan, Mark, "Ubiquity", Crown Publishers, NY 2001
22. Caes, Charles J., "Cosmology", Tab Books, Inc., Blue Ridge Summit, PA, 1986
23. Casti, John L., "Paradigms Lost", Avon Books, NY, 1989
24. Casti, John L., "Would be Worlds", John Wiley, 1997
25. Chown, Marcus, "The Quantum Zoo", Joseph Henry Press, 2006
26. Cole, K. C., "Mind over Matter", Harbour, 2001
27. Cole, K.C., "The Hole in the Universe", Harcourt, Inc., NY 2001
28. Davies, Paul, :About Time", Touchstone Books, 1995
29. Davies, Paul, :The 5th Miracle", Touchstone Book, London, 1999
30. Davies, Paul, "The Goldilocks Enigma", Mariner, 2006
31. Davies, Paul, "The Last 3 Minutes", Basic Books, 1994
32. Dawkins, Richard, "The God Delusion", First Mariner, 2008
33. Doidge, Norman, M.D., "The Brain that Changes Itself", Viking, 2007*
34. Drake, Stillman (Translator), "Discoveries & Opinions of Galileo", Randon House, 1957
35. Edelman, Gerard M., "Bright Air, Brilliant Fire", Basic Books, 2005
36. Ferguson, Kitty, "Measuring the Universe", Walker, 1999
37. Ferris, Timothy, "Coming of Age in the Milky Way", HarperCollins, NY, 1988
38. Ferris, Timothy, "The Whole Shebang", Simon & Sch., 1997
39. Flatow, Ira, "Present at the Future", HarperCollins. 2007
40. Freeman, Ken & McNamara, "In Search of Dark Matter", Praxis, 2006
41. Gingerich, Owen, "The Book Nobody Read", Penguin Books, England, 2004
42. Gleick, James, "CHAOS", Penquin Books, 1987
43. Gleiser, Marcelo, "The Dancing Universe", Plume, 1998

44. Goldblum, Naomi, "The Brain-Shaped Mind", Cambridge Press, UK, 2001
45. Goldsmith, Donald, "The Runaway Universe", Perseus, 2000
46. *Gott III, J. Richard, "Time Travel in Einstein's Universe", Houghton Mifflin Co., Boston, 2001*
47. *Green, Brian, "The Elegant Universe", Vintage Books, New York 1999*
48. Gribbin, John, "In Search of the Big Bang", Penguin Books, 1998
49. Guth, Alan, "The Inflationary Universe", Perseus, Cambridge, MA 1997
50. Hawking, Stephen and Mlodinow, Leonard, "A Briefer History of Time", Bantam Dell, New York, 2005
51. Hawking, Stephen, "Black Holes and Baby Universes and Other Essays", Bantam Books, London 1993
52. Hawking, Stephen, "The Universe in a Nutshell", Bantam Books, London 2001
53. Hawkins, Stephen, "A Brief History of Time", Bantam Books, London, 1988
54. Hawkins, Stephen, "On the Shoulders of Giants", Running Press, Philadelphia, PA, 2002
55. Heudin, Jean-Claude (Edited), "Virtual Worlds", Westview, 2004
56. Holmes, Hannah, "The Secret Life of Dust", John Wiley & Sons, 2001
57. Hooper, Judith, "Of Moths & Men", Norton, 2002
58. Kaku, Michio, "Beyond Einstein", Anchor Books, 1995
59. *Kane, Gordon, "Supersymmetry", Helix Books, Cambridge, MA 2000*
60. Klein, Etienne; Lachieze-Rey, Marc, "The Quest for Unity", Oxford University Press, 1999
61. Kragh, Helge, "Cosmology & Controversy, "Princeton Press, 1996
62. Laughlin, Robert B., "A Different Universe", Basic Books, New York, 2005
63. Lindley. David, "Uncertainty", Doubleday, 2007
64. Lloyd, Seth, "Programming the Universe", Alfred A. Knopf, 2006

65. Loewenstein, Werner R., "The Touchstone of Life", Oxford, 1999
66. Maddox, John, "What Remains to be Discovered", The Free Press, NY, 1998
67. Mazur, Joseph, "The Motion Paradox", Dutton, 2007
68. Miller, Arthur I., "Empire of the Stars", Hpughton Mifflin, 2005
69. Minsky, Marvin, "The Society of Mind", Simon & Schuster, 1985
70. Moffett, Shannon, "The Three-Pound Enigma", Algonquin, 2006
71. *Moring, Gary F., "The Complete Idiot's Guide to Theories of the Universe", ALPHA A Person Education Co., 2002*
72. Morrison, Philip & Phylis, "The Ring of Truth", Randon House, 1987
73. Ochoa, George; Hoffman, Jennifer; Tin, Tina; "Climate", Rodale, 2005
74. Park, David, "The Fire Within the Eye", Princeton University Press, 1997
75. Parker, Barry, "Creation-The Story of the Origin and Evolution of the Universe", Perseus Books, 1988
76. Penrose, Roger, " The Road to Reality", Alfred A. Knopf, 2004
77. Penrose, Roger, "The Emperor's New Mind", Oxford University Press, 1989
78. Pinker, Steve, "the blank slate", Viking, NY, 2002
79. Pollack, Robert, "The Missing Moment", Houghton Mifflin Co., 1999
80. Powell, James Lawrence, "Mysteries of Terra Firma", The Free Press, NY, 2001
81. Primack, Joel R., Abrams, Nancy Ellen, "The View from the Center of the Universe", Riverhead, 2006
82. Rees, Martin, "Just Six Numbers", Basic Books, 2000
83. Rizzi, Anthony, "The Science Before Science", IAP Press, Baton Rouge, LA 2004
84. Schwartz, Jeffrey M. and Begley, Sharon, "The Mind & The Brain", HarperCollins, NY 2002
85. *Siegfried, Tom, "The Bit and the Pendulum", John Wiley & Sons, Inc., New York 2000*

86. Silver, Brian, "The Ascent of Science", Solomon, 1998
87. Steinhardt, Paul J., Turok, Neil, "Endless Universe", Doubleday, 2007
88. Strogatz, Steven, "Sync", Hyperion, NY, 2003
89. Swimme, Brian & Berry, Thomas, "The Universe Story", Harper San Francisco,1992
90. Warshofsky, Fred, "Stealing Time", TV Books, 1999
91. Weinberg, Steven, "The First Three Minutes", Perseus Books, 1977
92. Wilson, Edward O., "The Future of Life", Vintage Books, 2002
93. Woit, Peter, "Not Even Wrong", Basic, 2006
94. Wolf, Fred Alan, "Parallel Universes", Touchstone, Simon & Schuster, NY, 1988
95. Yourgrau, Palle, "A World Without Time", Basic Books, 2005
96. Hawking, Stephen & Mlodinow, Leonard, "The Grand Design", Bantam, 2010
97. Carroll, Sean, "From Eternity to Here", Dutton, 2010

www.ingramcontent.com/pod-product-compliance
Lightning Source LLC
Chambersburg PA
CBHW022117170526
45157CB00004B/1677